YOUR KNOWLEDGE HAS VALUE

- We will publish your bachelor's and
  master's thesis, essays and papers

- Your own eBook and book -
  sold worldwide in all relevant shops

- Earn money with each sale

Upload your text at www.GRIN.com
and publish for free

Bibliographic information published by the German National Library:

The German National Library lists this publication in the National Bibliography; detailed bibliographic data are available on the Internet at http://dnb.dnb.de .

Imprint:

Copyright © 2018 GRIN Verlag
Print and binding: Books on Demand GmbH, Norderstedt Germany
ISBN: 9783668688261

This book at GRIN:

https://www.grin.com/document/421324

Hans Georg Schrey

Vacuum Drop Test of Air-Cooled Condensers in Operation

GRIN Verlag

Vacuum Drop Test of Air-Cooled Condensers in Operation

Author   Hans Georg Schrey

<u>Abstract</u>

Vacuum tightness is critical for air-cooled condensers operating at low absolute pressure. Low vacuum is aimed for because vacuum dominates the power plant efficiency. To verify vacuum tightness usually a vacuum drop test is made with the system empty at normal atmospheric temperature and free of any liquids. This test is done before commissioning of the power unit and generally follows the recommendations of the Heat Exchanger Institute (HEI) as outlined in §6.1.1 of "Standards for Steam Jet Vacuum Systems".

However, over time of operation the power plant may develop leakages, which were not present at the time of the original drop test. This calls for a tightness test at operating conditions where pre-conditions for the standard vacuum drop test are not fulfilled. The report describes a vacuum drop test without interfering too much into normal power plant operation. The test is suitable for stationary operating conditions using standard operation readings.

The assessment of leakage flow is based on the measured vacuum decay rate. It is shown that vacuum decay rates taken from tests before and after commissioning are different. Contractual fixing of acceptable vacuum decay rates should therefore be treated with care. Example graphs for easy evaluation are given.

Keywords: Air-cooled Condenser; ACC; Air Leakage; Vacuum decay; HEI; VGB-R126L; VGB-R131

Table of contents

1 Introduction

To safeguard stable operation all vacuum systems must be hermetically sealed. Therefore, prior to commissioning of the vacuum system the vacuum tightness is checked [1, 2]. Generally, these tests are made according to the recommendations of the Heat Exchanger Institute, HEI [3]. Air-cooled condensers ("ACC") are characterized by a huge vacuum volume which is due in the first place to large steam ducts. The condenser bundles constitute only a small fraction of the overall volume. A short description of ACC features is shown in fig. 1 below.

A typical ACC consists of one or more parallel streets with a multitude of heat exchangers arranged in an A-frame geometry at approximately 60° base angle. The condensation is done in two steps with primary and secondary condenser in sequence. The tube bundles are combined in individual heat exchanger modules which are similar in geometry. The set of heat exchangers forming a module is normally served by one fan. Steam supplied via the turbine exhaust duct enters the primary condenser section at the exchanger top manifold and flows down the tubes to the condensate collection line ("CCL"). A considerable fraction of steam is condensed along the flow path through the primary condenser and, at the same time undergoes a pressure drop due to friction. Condensate and residual steam are collected in the CCL which conveys the remaining steam to the secondary condenser section ("dephlegmator") and all condensate via the condensate line to the condensate tank. At dephlegmator inlet the remaining steam enters from below and is condensed during up-flow in the exchanger tubes. The secondary condensate is drained back to the CCL by gravity and re-heated to CCL temperature. All inert gases are carried to the dephlegmator outlet along with the steam flow. The dephlegmator exit connects to the inert gas evacuation unit which keeps the vacuum permanent. To suppress extreme condensate sub-cooling the tank may be re-heated via an

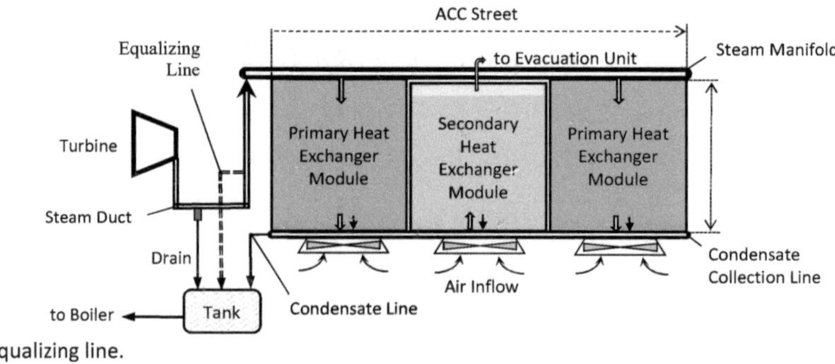

equalizing line.

Fig. 1: General ACC

2 Standard Vacuum Drop Test

The standard vacuum drop test procedure used for ACC's is described in "Standards for Steam Jet Vacuum Systems", §6.1.1 issued by HEI [3]. The test is done with the ACC out of operation. The ACC is treated as static volume losing its vacuum through leakages which allow influx of ambient air into the system. To perform the test the steam duct inlet must be sealed and the total vacuum volume properly assessed. However, leakages in the vacuum range of the turbine may not be included. The absolute leakage rate is paramount for accurate design of the evacuation unit.

With a test over a short period of time at constant ambient temperature and constant total volume the ideal gas equation of state holds:

$$\frac{d(p\,V)}{d\tau} = R_A\,T_A \cdot \frac{dM}{d\tau} \tag{1}$$

The change of mass per time is same as the inflow of inert gas (or leakage rate)

$$\mu_{stat} = \frac{dM}{d\tau}$$

with the vacuum decay rate

$$\dot{p}_{stat} \cong \frac{p_1 - p_0}{\Delta\tau}$$

Summarizing, $\quad \mu_{stat}(\tau) = \frac{V}{R_A\,T_A} \cdot \dot{p}_{stat}$ $\tag{2}$

This leads for 20°C ambient temperature to the following quantity equation:

$$\frac{\mu_{stat}(\tau)}{[kg/h]} \cong \frac{1}{14{,}08} \cdot \frac{V}{[m^3]} \cdot \frac{\dot{p}_{stat}}{[mbar/min]}$$

which corresponds to the formula given in §6.1.1 of [3]. Therefore, to estimate the leakage rate at static conditions only vacuum decay rate and total volume are needed. Figure 2 shows example graphs of equation (2) as function of the vacuum decay rate for ambient temperatures in the range of 5°C to 35°C. The lower the ambient temperature the higher the leakage rate.

In practical applications it is common to use the vacuum decay rate as leakage indicator (see fig. 3). This value is based on HEI recommendation of inert gas flow rate in vacuum condensers [4]. The ACC supplier uses these data to properly design the evacuation system capacity. The examples of vacuum decay rate in figure 3 have been calculated as function of the total ACC volume for 20°C ambient temperature. As could be expected, the smaller the ACC volume the higher the vacuum decay rate.

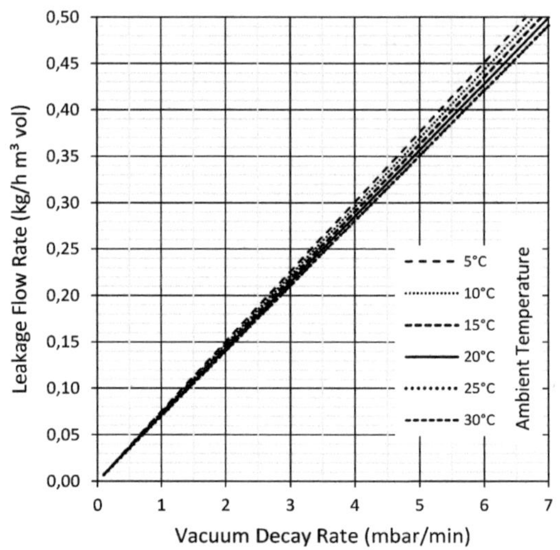

Fig. 2: Static HEI Leakage Test

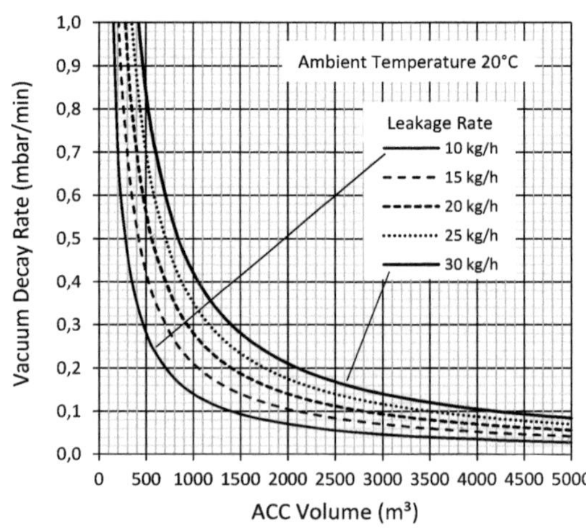

Fig. 3: Static Vacuum Decay Rate, 20°C

3 Leakage Test with ACC in Operation

With the ACC in operation the pre-conditions for using the HEI vacuum test are no longer applicable [3]. Therefore, §3.3.5 of VGB ACC acceptance test standard [1] recommends an air leakage test following VGB Guideline R-126 L ("Recommendations for Design and Operation of Vacuum Pumps for Steam Turbine Condensers") [5]. However, this test requires a multitude of single measurements and the modification of the vacuum piping by different standardized leakage nozzles. So, normal operation of the ACC is somehow affected during the test. Therefore, at time of erection the HEI test is accepted by clients and suppliers of ACC equipment as confirmation of vacuum tightness in deviation from the VGB test code [1].

None-the-less a situation may arise after commissioning of the power unit to check the vacuum tightness of the system because of loss of vacuum. A leakage test avoiding extra time and effort by using operational data would be extremely desirable. For the proposed simplified method we need to consider at least steady state operation in order to come close to meaningful results.

For the following procedure the steam saturation line is needed. A good approximation is

$$t_s(p) = \frac{a}{[b - \ln(p)]} + c$$

or
$$p_s(t) = exp\left[b - \frac{a}{t-c}\right]$$

with $a = 3888{,}11$; $b = 23{,}3084$; $c = 43{,}2$ (SI units).

For the ACC operation decay test the inert gas evacuation is blocked for some time (by closing the evacuation valve), while all operational parameters (steam flow, fan settings, ambient conditions) are kept constant. The loss of vacuum due to inert gas accumulation is visible by the change of turbine exhaust pressure over time. This suggests that the HEI method may apply because again, the ACC volume is filled over time with inert gas. However, accumulation of inert gas impedes also the heat exchange process which has an effect on condensation pressure level as well.

Therefore, we have two sources of vacuum decay during ACC operation:

   (a) Inert gas accumulation in ACC, Δp_a

   (b) Loss of condensing capacity caused by blanketing of effective surface, Δp_b

For simplification we estimate the vacuum decay of each effect separately. The higher value controls the overall pressure increase. The turbine exhaust pressure dominates the thermohydraulic state of the ACC and is therefore selected as chief performance parameter. Thus, the effective vacuum decay is

$$\Delta p_{EX} = max\left\{\Delta p_{EX,a}; \Delta p_{EX,b}\right\} \tag{3}$$

3.1 Inert Gas Accumulation in ACC during Operation

The inert gas filling of the ACC volume follows the same law as the static volume. Over the test period there will be a small change of temperature. All inerts are quickly heated up to steam temperature. The terminal steam temperature level is crucial for the terminal gas content. As only traces of inerts are contained in the main steam flow the properties of steam will be good enough as approximation for the mixture. Equation (1) therefore turns into

$$\Delta p_{EX,a} = \frac{R_S\, t_{EX,1}}{V} \cdot \mu_{op} \cdot \Delta\tau \tag{4}$$

This differs from the HEI model (1) by a factor of $(R_S\, t_{EX,1})/(R_A\, T_A)$. Note that equation (4) takes the form of equation (2) if we exclude condensation and skip to air fill.

3.2 Loss of Condensing Capacity Caused by Surface Blanketing

For the purpose of this investigation the general ACC layout (fig. 1) will be simplified to some extent. The ACC is considered as volume reaching from turbine exhaust to dephlegmator outlet. Steam entering the ACC at turbine exhaust carries all inerts along the flow path to the dephlegmator exit where all steam has been consumed by condensation. Figure 4 shows the simplified system with primary and secondary condenser. The following parameters shall be available from operation records:

- Elapsed time for the test
- Turbine exhaust pressure at begin and end of test
- Condensate temperature, measured at CCL level before tank
- Steam flow rate
- Ambient temperature
- Fan speed settings

Only the first three parameters are of significance as the test will be performed under constant operating condition (flow rate, ambient temperature, fan speed ...)

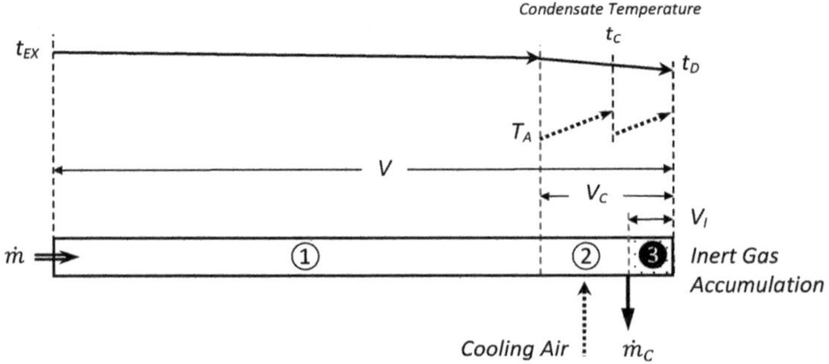

Fig. 4: Simplified ACC Model

As mentioned, turbine exhaust parameters are pre-dominant for steam flow as well as condensation duty. In the range of the ACC vacuum zone three sections may be discerned (see fig. 4): ① initial steam duct supplying steam to the condensers with no condensation, ② bundle range with normal condensation comprising primary condenser and initial part of the dephlegmator, and ③ top region of dephlegmator bundles with accumulation of all inert gas (region of surface blanketing).

The most important region is section ③. Initially, all sections are filled with steam and only traces of inert gas. This means that inert partial pressure is so low that steam side temperature and absolute pressure are linked by the steam saturation line. At stop of the evacuation the inert gas conveyed with the steam is no longer withdrawn from the dephlegmator outlet and resides there permanently. This goes along with a drop of local temperature which quickly reaches ambient cooling level. The surface blanketed by inert gas-steam mixture is lost for heat transfer and does no longer contribute to condensation. With constant steam flow under steady state operation the system makes up for the reduction of effective surface by raising the initial temperature difference. This leads to an increase of exhaust steam pressure or, loss of vacuum. Thus, even a small air leakage may result in noticeable increase of system pressure. Over time, the active dephlegmator section in ② is gradually reduced by the blanketed volume. At dephlegmator exit in section ③ we find a steam-air mixture saturated at ambient air temperature.

Vacuum decay caused by loss of cooling can be deduced from the heat balance

$$\left(\frac{\dot{m}}{\dot{m}_0}\right) \cdot \left(\frac{r}{r_0}\right) \cdot \left(\frac{x_{EX}}{x_{EX,0}}\right) = \left(\frac{\rho_A \cdot c_P \cdot u}{\rho_{A,0} \cdot c_{P,0} \cdot u_0}\right) \cdot \left(\frac{A}{A_0}\right) \cdot \left(\frac{\Phi}{\Phi_0}\right) \cdot \left(\frac{\vartheta}{\vartheta_0}\right).$$

With constant operation parameters the exchanger effectiveness of the active surface is approximately constant as well. Also, the change of condensation heat and steam wetness is small and may be neglected. Thus, we find for active face area

$$\left(\frac{A}{A_0}\right) = \left(\frac{\vartheta}{\vartheta_0}\right)^{-1} .$$

Steam encounters a pressure drop along the flow path through ducts and tubes which in turn, reduces the local steam temperature (fig. 4). Normally, the temperature change is much smaller on the steam side than on the air side. Consequently, the heat exchange process is close to isothermal. Considering duct pressure drop as an abatement of exchanger effectiveness we simply take the turbine exhaust temperature as controlling the initial temperature difference:

$$\frac{\vartheta}{\vartheta_0} = \frac{t_{EX,1} - T_A}{t_{EX,0} - T_A} = \frac{t_s(p_{EX,1}) - T_A}{t_s(p_{EX,0}) - T_A}$$

The loss of condensation duty is caused by the blanketed terminal zone of the dephlegmator. Steam and air residing there no longer contribute to heat exchange after being cooled down to ambient level. So, loss of effective surface goes along with formation of an inert gas-steam mixture, so that

$$\left(\frac{A}{A_0}\right) = 1 - \frac{V_I(\tau)}{V_C}$$

or,
$$\frac{V_I(\tau)}{V_C} = \left[1 - \left(\frac{\vartheta}{\vartheta_0}\right)^{-1}\right] \tag{5}$$

The blanketed volume contains a steam-air mixture saturated at ambient temperature. The water content of this mixture is

$$x = 0{,}622 \cdot \frac{p_s(T_A)}{p_{D,1} - p_s(T_A)}$$

and the affected volume

$$V_I = m_L \cdot \frac{R_A \, T_A}{p_{D,1}} \cdot (1 + 1{,}607 \, x)$$

Combining these equations we find the blanketed volume as

$$V_I(\tau) = \frac{R_A \, T_A}{p_{D,1} - p_s(T_A)} \cdot \mu_{op} \cdot \Delta\tau \tag{6}$$

Normally, the dephlegmator exit pressure needed in equation (6) is not recorded. It must therefore be derived from temperature readings. Note that the dephlegmator exit (or evacuation line) temperature cannot be used to calculate the absolute pressure by means of the steam saturation line because of accumulation of an unknown fraction of inert gas. A better indication of state is the condensate temperature at CCL level. Typically, condensate sub-cooling is restricted to the low Kelvin range:

$$\Delta t_{sub,C} = t_s(p_{EX}) - t_C < 5 \, K$$

Condensate sub-cooling should be as low as possible. Values around $2 \, K$ are standard in steam cycle design. Condensing steam flowing in the dephlegmator undergoes some additional sub-cooling which is in the range of $\Delta t_{add} \approx 1 \, K$ – so that in total

$$\Delta t_{sub} = (\Delta t_{sub,C} + \Delta t_{add})$$

Dephlegmator exit temperature may now easily be estimated as $t_{D,1} = t_{EX,1} - \Delta t_{sub}$ leading to an average dephlegmator exit pressure of $p_{D,1} = p_s(t_{D,1})$.

To estimate the steam pressure effect the right-hand-side of equation (5) is evaluated. Considering only small changes of operating parameters we may linearize initial temperature difference and steam saturation line to

$$\frac{p_{EX,1}}{p_{EX,0}} - 1 \cong \frac{\vartheta_0}{S} \cdot \left[\left(\frac{\vartheta}{\vartheta_0}\right) - 1\right] \cong \frac{\vartheta_0}{S} \cdot \left[1 - \left(\frac{\vartheta}{\vartheta_0}\right)^{-1}\right] = \frac{\vartheta_0}{S} \cdot \frac{V_I(\tau)}{V_C}$$

with
$$S \equiv \frac{a}{[b - \ln\{p_{EX,0}\}]^2} = \frac{[t_{EX,0} - c]^2}{a}$$

Consequently, $\quad \Delta p_{EX,b} = p_{EX,0} \cdot \left(\frac{p_{EX,1}}{p_{EX,0}} - 1\right) = p_{EX,0} \cdot \frac{\vartheta_0}{S} \cdot \frac{V_I(\tau)}{V_C} \tag{7}$

3.3 Effective Pressure Effect

The total pressure effect is according to equation (3) (by using equations 6 and 7):

$$p_{EX,1} - p_{EX,0} = max\left\{1{,}607 \cdot \frac{t_{EX,1}}{T_A}; \ \frac{p_{EX,0}}{p_{D,1} - p_s(T_A)} \cdot \frac{\vartheta_0}{\alpha \, S}\right\} \cdot \frac{R_A \, T_A}{V} \cdot \mu_{op} \, \Delta\tau$$

The measured vacuum decay rate is

$$\dot{p}_{op} \cong \frac{p_{EX,1} - p_{EX,0}}{\Delta\tau}$$

so that

$$f_{op} \cdot \mu_{op} = \frac{V}{R_A \, T_A} \cdot \dot{p}_{op} \tag{8}$$

with

$$f_{op} = max\left\{1{,}607 \cdot \frac{t_{EX,1}}{T_A}; \ \frac{1}{\alpha \cdot f}\right\} \tag{9}$$

and

$$f = \frac{p_{D,1} - p_s(T_A)}{p_{EX,1} - p_{EX,0}} \cdot \frac{t_{EX,1} - t_{EX,0}}{t_{EX,1} - T_A} \tag{10}$$

In equation (10) the following equivalence was used:

$$S \cong p_{EX,0} \cdot \frac{t_{EX,1} - t_{EX,0}}{p_{EX,1} - p_{EX,0}}$$

The blanketing effect is restricted to the exchanger volume fraction of the total ACC. Loss of active surface will only materialize if the peak steam pressure is below ambient air pressure. Following the recommendation of HEI the steam pressure should never exceed 50 kPa (i.e. approx. 80°C).

For ease of overview the operation pressure factor $'f'$ which subsumes all steam side operation parameters has been illustrated in figures 5 to 7. Looking at the figures we see that the lower the sub-cooling level the higher the operation pressure factor $'f'$ (i.e. leakage rate) will be. This factor also rises with the absolute vacuum decay (lines between 0,1 and 5 kPa are shown). Last not least, the lower the start pressure the higher is the operation pressure factor.

Note that in all graphs $f < 1$. This is intensified by the fact that – in most cases - the exchanger volume fraction is only 20% or less of the total ACC. A product of $\alpha \cdot f = 10\%$ or less may easily result. In terms of f_{op} (equation 9) the reciprocal summand is around 10 or higher and, thus larger than the static accumulation term which is around 1,7. In consequence, the surface blanketing term is controlling equation (9) and therefore will be used in general. The accumulation effect plays only an inferior role for inter-dependency of leakage rate and vacuum decay for ACC in operation.

Theoretically, if steam temperature approaches ambient temperature sub-cooling will be annihilated and the operating pressure factor approaches $f \to 1$. This will only happen if the condensation duty is heavily reduced from design conditions. The restriction to exchanger volume fraction < 1 makes no more sense in this situation. Therefore, with diminishing heat exchange $\alpha \to 1$ as well. However, with $f_{op} \to 1$ we find the static HEI solution as limit value of equation (9).

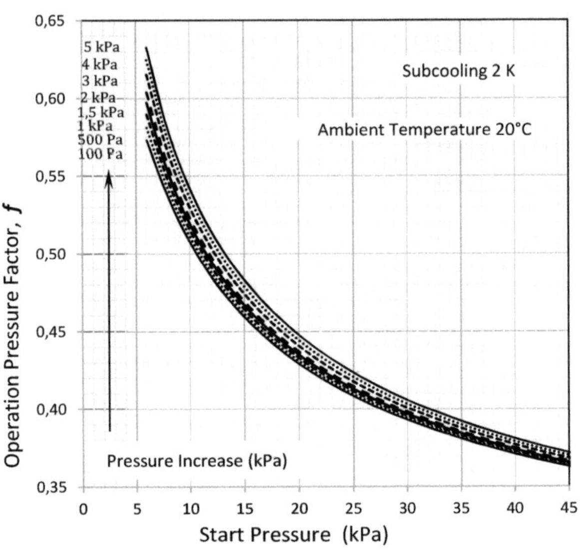

Fig. 5: Operating Pressure Factor, 2 K Sub-cooling

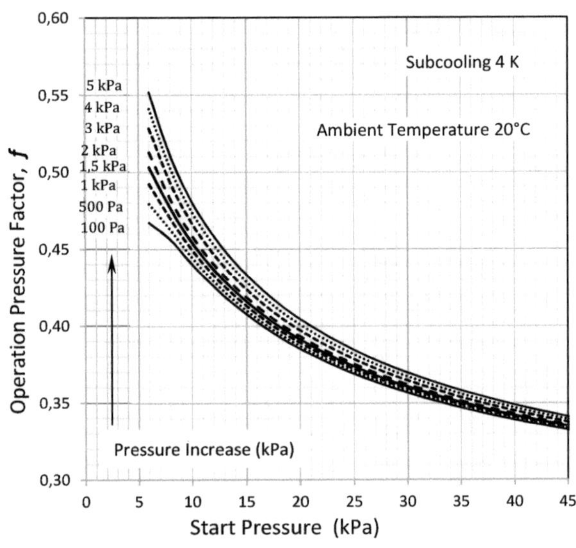

Fig. 6: Operating Pressure Factor, 4 K Sub-cooling

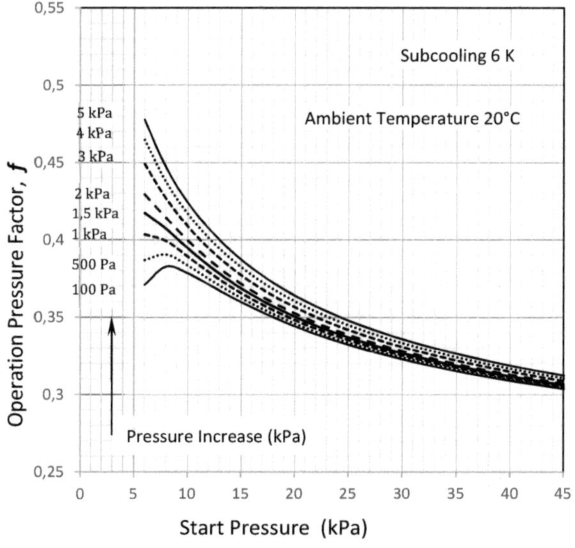

Fig. 7: Operating Pressure Factor, 6 K Sub-cooling

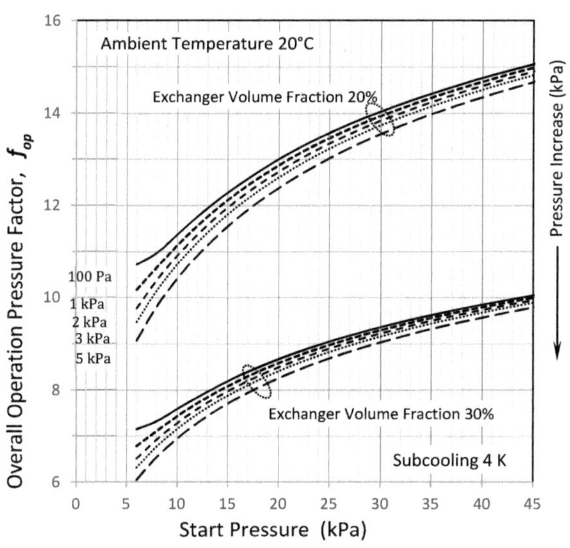

Fig. 8: Overall Vacuum Decay Enlargement Factor

4 Comparison of Static and Steady State Results

Normally, vacuum tightness is checked by measuring the vacuum decay rate and comparing to a pre-assessed value. If the design leakage rate μ_{des} is known [4] the acceptable vacuum decay rate may be found. To this end equations (2) and (8) must be re-arranged.

$$\dot{p}_{stat} = \frac{R_A T_A}{V} \cdot \mu_{des} \tag{2a}$$

$$\dot{p}_{op} = f_{op} \cdot \frac{R_A T_A}{V} \cdot \mu_{des} = f_{op} \cdot \dot{p}_{stat} \tag{8a}$$

For clarity we select a typical sub-cooling value of 4 K as an example for the overall operation pressure factor, f_{op}. Figure 8 shows curves for 20% and 30% exchanger volume fraction. Curves have been created for a supposed pressure increase between 0,1 and 5 kPa during the test.

It is easy to see that basically, the vacuum decay rate is at least a magnitude greater in operation than the static HEI value. We also see the massive impact of exchanger volume fraction. As could be expected – larger volume fraction leads to lower enlargement factors. Also, the lower the steam test pressure the lower the enlargement factor will be.

If we take, as an example the following data: ACC volume $V = 2000\ m^3$, leakage $\mu_{des} = 30\ kg/h$ and ambient temperature 20°C we find from figure 3 the allowable static vacuum decay rate of

$$\dot{p}_{stat} = 0,21\ mbar/min.$$

The operation test may have the following pressure at begin of the test $p_{EX,0} = 15\ kPa$. At the end of the test the pressure may be increased by $\Delta p_{EX} = 1\ kPa$, with a sub-cooling of $\Delta t_{sub} = 4\ K$. Figure 8 shows for an exchanger volume fraction of 20% $f_{op} \approx 12{,}1$. This means that pressure decay rate of ACC in operation, $\dot{p}_{op} = 2,54\ mbar/min$ still confirms the nominal leakage flow.

5 Conclusion

In spite of the thermohydraulic complexity of the problem it is possible to assess leakage rates based on control room readings. Although the proposed procedure is based on the heat exchange process it does not need specific design point data or details of the fin tube system. The only design figures required are total and exchanger volume.

As expected, the static drop test result as described in HEI standard is not applicable for steady state ACC operation. An alternative method must be used, such as the proposed simplified procedure.

A fixed vacuum decay rate for all conditions cannot be defined. The results show that static and steady state leakage rates differ by at least one magnitude for same vacuum decay rate.

Glossary

$\alpha = \frac{V_C}{V}$ -   heat exchanger volume fraction of total ACC vacuum volume

ϑ -   initial temperature difference: steam - air (°K)

Φ -   heat exchanger effectiveness (-)

$\tau, \Delta\tau$ -   time variable, elapsed time (s)

μ_{des} -   acceptable leakage mass flow (kg/s), design value

μ_{op} -   average leakage mass flow (kg/s), ACC in operation

μ_{stat} -   average leakage mass flow (kg/s), static HEI test

a, b, c -   steam saturation curve constants

A -   active air side face area of exchangers (m²)

c_P -   air specific heat capacity (J/kg K)

f -   operating pressure factor (-)

f_{op} -   overall vacuum decay enlargement factor (-)

m_L -   mass of leaked air contained in ACC (kg)

M -   total mass in ACC (kg)

p_D -   dephlegmator steam pressure (Pa)

p_{EX} -   turbine exit steam pressure (Pa)

$\Delta p_{EX} = p_{EX,1} - p_{EX,0}$ -   total vacuum decay (Pa)

$\Delta p_{EX,a}$ -   vacuum decay due to gas accumulation (Pa)

$\Delta p_{EX,b}$ -   vacuum decay due to blanketing of active surface (Pa)

$p_s(t)$ -   steam saturation pressure (Pa) at temperature 't'

$\dot{p}_{op} = \frac{\Delta p_{EX}}{\Delta\tau}$ -   vacuum decay rate (Pa/s), ACC in operation

$\dot{p}_{stat} = \frac{\Delta p_{EX}}{\Delta\tau}$ -   vacuum decay rate (Pa/s), HEI test

r -   condensation heat (J/kg)

R_A -   dry air gas constant = 287,1 (J/kg K)

R_S -   steam gas constant = 461,4 (J/kg K)

S -   auxiliary group for saturation line

t -   steam temperature (K)

t_C -   condensate temperature (K)

t_D -   temperature in sub-cooled dephlegmator volume (K)

T_A -   ambient air temperature (K)

t_{EX} -   turbine exhaust steam temperature (K)

$t_s(p)$ -   steam saturation temperature (K) at pressure 'p'

Δt_{add} -   additional sub-cooling of dephlegmator inlet to top region (K)

$\Delta t_{sub,C}$	-	sub-cooling (K) = difference of steam and condensate temperature
Δt_{sub}	-	total sub-cooling between turbine exit and dephlegmator end (K)
u	-	air face velocity (m/s)
V	-	total ACC vacuum volume (ducts, heat exchangers, etc.) (m³)
$V_C = \alpha \cdot V$	-	heat exchanger volume (m³)
V_I	-	blanketed volume (m³)
x	-	water content (kg/kg) of dry air
x_{EX}	-	steam quality (kg/kg) at turbine exhaust

<u>Indices</u>

| 0 | - | start of test period |
| 1 | - | end of test period |

Bibliography

[1]   "VGB Guideline Acceptance Test Measurements and Operation Monitoring of Air-Cooled Condensers under Vacuum", VGB R-131 Me, VGB Kraftwerkstechnik GmbH, Essen, 1997

[2]   "Air-Cooled Steam Condensers, Performance Test Codes", ASME PTC 30.1 - 2007, The American Society of Mechanical Engineers, 2008

[3]   "Standards for Steam Jet Vacuum Systems", 5th edition 2000, Heat Exchange Institute, Cleveland

[4]   "Standards for Air Cooled Condensers", 1st edition 2011, Heat Exchange Institute, Cleveland

[5]   "Empfehlung für Auslegung und Betrieb von Vakuumpumpen bei Dampfturbinen-Kondensatoren" (Recommendation for Design and Operation of Vacuum Pumps for Steam Turbine Condensers) VBG R-126 L, VGB Kraftwerkstechnik GmbH, Essen (in German)